THEORY OF ALMOST EVERYTHING

NEW CALCULUS

UNIVERSE WITHOUT DARK MATTER OR DARK ENERGY

GRAVITATION WITHOUT ANY CONSTANTS

UNIFICATION OF GRAVITY AND ELECTROMAGNETISM

SCORE THEORY

BY VAIBHAV JAIN

A time interval is always continuous, no matter how small, there can be no smallest possible time interval as any you take will be made up of smaller components. There can only be one smallest possible time interval that is zero. Same is true for any continuously changing variable.

While using differentiation for finding or expressing the laws of the universe we always need the 'first cause' and the effect and for finding the initial cause of an event we always need 'the smallest possible time interval an extremely small one won't do.

Suppose we are differentiating a variable x with respect to time, we first need to take the change in x in the smallest possible time interval, which has to be zero, meanwhile in an extremely small time interval x becomes x + Δx but we ignore the Δx and take dx to be zero(as dt is zero). In this way if we keep on ignoring Δx the difference between actual value of x ie x + Δx , and book value of x keeps growing ,now suppose x + Δx is multiplied by another variable (or even constant) the difference between x + Δx and x increases more rapidly. Suppose x is a constantly changing variable then at the beginning when x's value was Δx, here comes our first rule ,none of our chosen variables(namely distance, velocity, cos a, sin a) can ever be zero, so when x was Δx and increased to 2Δx, then we ignored 100% of increase in x, now when x changes from 2Δx to 3Δx we again ignore Δx ,ie an increase of 50% of x ,when x is 3Δx and changes to 4Δx we ignore 33.33 % of x ,when x is x and increases to x+ Δx ,we ignore an increase of (Δx/x)*100% of x, in this way we ignore 100% +50% +33.33% +25%... approximately lnx*100% of x . So while the book value is x the actual value is x(1+lnx) ,the book value is such because we have taken dx instead of Δx at all intervals ,but then why take percentage ,why not just add up and conclude that we have taken 0 instead of x. That is because when x is multiplied with other numbers the difference also grows larger at each step ,but what when something is added to x, then the difference doesn't grow larger ,well I will explain that case later.

To differentiate a variable we will first have to convert the book value of the variable into actual value

x = x(1+lnx)

Now dx = 0 but d(lnx) is not = 0 ,You may say to differentiate lnx we again need to convert it into actual value, and it will be an endless cycle but over here we need the 'change in x' in the smallest possible interval 'the percentage of x we are ignoring' can vary and we are under no constrain to 'ignore the smallest possible percentage of x' at each step in other words Δx can vary and Δx does not need to be measured in the smallest possible time interval ,a reasonably small one will do.

d(x(1+lnx))/dt= 0(1+lnx) + xd(1+lnx)

$$=0 +x*(dx/dt)/x$$

But the x above is x at book value(x_b) and x at denominator is x at actual value

$=(dx/dt)/(1+lnx)$

This is the differential of x. Over here dx/dt is actually $\Delta x/\Delta t$ where Δx and Δt are extremely small(but not zero). We can rewrite dx/dt as $\Delta x/\Delta t(1+lnx)$. Now before taking ln of x we must convert physical quantities into their natural units ,we cannot take just any unit. Natural unit of velocity is c, that of distance is $1.2*10^{-25}$meters. This is necessary as we won't be using any constants except c (and h once to derive natural units of distance, apart from that only measured masses and distances between things)

Although we refer to x(1+lnx) as actual value ,for calculations it will always have to be written as x(1+lnx) and never as only x and the distinction between actual value and book value will come into play only when we differentiate a continuously changing variable in other words x will become x(1+lnx) only before we differentiate a continuously changing variable x.

The differential of 1/x will be

$1/x=(1/x)_b (1+ln(1/x))$

$=(1/x)_b (1-lnx)$

$=d((1/x)_b(1-lnx))/dt$

$=(1/x)_b\Delta(1-lnx)/\Delta t$

$=(1/x)_b*(-\Delta x/x)$

$=-(\Delta x(1+ln(1/x))/x)_b*(1/x)_b$

$=-\Delta x(1-lnx)/x*x$ as x_b is written simply as x while x as x(1+lnx)

Now if we want to differentiate xy

$xy = (xy)_b (1+\ln(xy))$

$= x_b y_b (1+\ln x + \ln y)$

$= x_b y_b \Delta(1+\ln x + \ln y)/\Delta t$

$= x_b y_b (\Delta x/x + \Delta y/y)$

$= \Delta xy/(1+\ln x) + \Delta yx/(1+\ln y)$

Now you may ask why not take xy as one variable and get the differential as $\Delta(xy)/(1+\ln(xy))$, because, here comes the second rule-

You must separarate the ln of a term into ln of independent variables before differentiating, an independent variable may be defined as one which does not neccecarily change when some other variable changes and whose slope at a point(ie an extremely small space) is the least curved.

Now if you have to differentiate √x

$\sqrt{x} = \sqrt{x_b} (1+\ln\sqrt{x})$, now since √x is not an indepent variable we must convert the ln of it into ln of the independent variable

$= \sqrt{x_b} (1+(\ln x)/2)$

$d\sqrt{x_b}(1+(\ln x)/2)/dt$

$= \sqrt{x_b}\Delta(1+(\ln x)/2)/\Delta t$

$= \sqrt{x_b} (\Delta x/2x)$

$= \Delta x/2\sqrt{x_b}(1+\ln x)$

$= \Delta x/2\sqrt{x}(1+\ln x)$

If a variable x is written as x/2+x/2 and then differentiated

The differential will be $(\Delta x/2)/(1+\ln x) + (\Delta x/2)/(1+\ln x)$ it will not become $(\Delta x/2)/(1+\ln(x/2)) + (\Delta x/2)/(1+\ln(x/2))$

Suppose we can write x as y+z ,now if we differentiate x and supposing y and z are independent ,the differential will be $\Delta y/(1+\ln y) + \Delta z/(1+\ln z)$ the fact that z has been added to y makes no difference to lny but if x were=y*z and now even if y and z are independent ,the differential will be $z*\Delta y/(1+\ln y) + y*\Delta z/(1+\ln z)$ and as you can see lny is part denominator of a much larger number as along with the differential of y it(lny) has also been multiplied with z ,so in multiplication the distinction between book value an actual value grows while in addition it does not and that is why while finding the actual value of a continuously changing variable we took percentage at each step.

The differential has changed but the integral will remain the same as traditional calculus because during integration we just sum up extremely small or even smallest possible changes.

$\int \Delta x dx/(1+\ln x) = x/(1+\ln x) - 1/(1+\ln x)^2 \ldots$

PART   2

The universe is infinite in space and continuously evolving. In this dynamic universe many systems exist, each of them faces 2 risks ,firstly due to circumstances it might change so much that the new system has no resemblance whatsoever to the previous system and the previous system may become history or in other words die. Second risk a system faces is that in this infinite universe it may get so far away from other systems that it may have no effect on any other system in other words it may get lost and isolated in space. Let us assume this universe is made up of empty spaces and tiny particles, then any system can be defined as the number of particles and the spaces between them. The best system will be the one that is changing the least (at any given period of time) and is closest to other systems. Another factor to consider will be that if 2 particles separated by 10000 meter start moving away at 1 cm per second ,we will call that change as slow while if 2 particles separated by 0.00001 cm start moving away at 1 cm per second we will call that change as fast.

Let's go step by step

A system that changes the least is the best

1-(vcosa)^2/2 where v stands for velocity of a particleand cosa for the cosine of direction of same particle with respect to some other particle.

A system that is closest to other to other systems is the best.

(1/r)-(vcosa)^2/2 where r stands for distance from 1 other particle

Considering the third factor

(1/r)-((vcosa)^2/2r)

Now the vcosa term will be multiplied by the mass of the particle ,the relativistic mass .For now we will assume all particles have equal and unit mass.

(1/r)-((vcosa)^2/2r(√(1-v^2/c^2))

This is the score of a particle with respect to one other particle. Total score of a particle is the sum of it's score with respect to all particles.

$$= (1/r)-((vcosa)^2/2r(\sqrt{(1-v^2/c^2)})) + (1/r2)-((v2cosb)^2/2r(\sqrt{(1-v2^2/c^2)})) + (1/r3)-((v3cosc)^2/2r3(\sqrt{(1-v3^2/c^2)}))...$$

Where v2 ,r2 ,cosb are velocity distance and direction with respect to a second particle and v3,r3,cosc with respect to a third particle. Each 1/r term will be multiplied by respective mass of the corresponding particle.

If most of other particles are concentrated at a far away point, score of a particle can be approximated as

Ng/Rg −Ng(vcosa)^2/2Rg(y) where Ng stands for mass of other particles in the galaxy/galaxy cluster and Rg your distance from the point where other particles are concentrated ,v is your velocity with respect to that point and cosa the cosine of your angle from that point and y the relativistic term. But this equation must be written in natural units. If v refers to velocity in meter per second ,velocity in natural units become v/c.

The equation now becomes

Ng/Rg −Ng(vcosa)^2/2(c)^2Rg(y) where c stands for speed of light in vaccum.

This: Ng/Rg −Ng(vcosa)^2/2(c)^2Rg(y)=score of a particle

Score of all particles summed up is the score of the universe and the score of the universe is always conserved. The score cannot increase to infinity, who would want to live in a black hole anyway. In fact the score of the universe cannot increase (or decrease) at all, in this universe you can improve only at the cost of others that's why its evolving, only the fit survive.

The universe is effectively divided into galaxy, galaxy clusters and super clusters with most mass being centered at the center of galaxy, galaxy cluster and super cluster (ignoring the dark matter which trust me doesn't exist). The distance between two super clusters is huge and score from other super clusters will get too small for particles in one super cluster and can be effectively ignored. While analyzing the motion of a collection of particles that move together (a body) we will have to ignore the score of constituent particles from each other because no score disturbance takes place over there and even if there is relative motion between the particles the body's movement is independent of it.

More than 99% of the mass of a galaxy is concentrated at the center of the galaxy (again if we assume dark matter does not exist). And most of the mass of galaxy super cluster exists at the center of the super cluster (at least in the case of Virgo super cluster if not always). For particles which are not too close to a large body the $1/r + 1/r2 + 1/r3$ .. part of its score can be approximated as $(Ng/Rg + Ngs/Rgs)$ where Ng is mass of particles in the central region of the galaxy ang Rg, the chosen particles distance from it and Ngs mass of particles in the central region of galaxy super cluster and Rgs the chosen particle's distance from it. If unlike local cluster which is a poor galaxy cluster if a galaxy cluster is proportionately massive enough we will also have to take the number of particles in galaxy cluster center in addition to galaxy center and galaxy super cluster center.

Now let us start by analyzing the motion of a star around the center of its galaxy. For sake of simplicity let's assume it's the sole galaxy in the universe (no, no cluster etc) and almost all its mass is concentrated in the center. Then the score of the star will be

$nNg/Rg - nNg(vcosa)^2/2Rg(c^2)(y)$ where v is the velocity with respect to the center of galaxy, cosa the cosine of angle it makes from the center of galaxy and n the number of particles in the star. Unless some other particle compensates for change in score of a particle at the exact same time the score a particle remains constant. The score of the star which is sum of scores of constituent particles will also be constant.

Therefore we have

$nNg/Rg - nNg(vcosa)^2/2Rg(c^2)(y) = constant$

Differentiating both sides with respect to time (we no longer need to write further derived quantities in natural units except the units with ln in front of them)

⇨ $-nNg(1-lnRg)vcosa/Rg^2 - nNgvA(cosa)^2/Rg(c^2)(y)(1+lnv) +$
$nNgv^3cosa(sina)^2/(Rg^2)(c^2)(y)(1+lncosa) +$
$nNg(vcosa)^3(1-lnRg)/(Rg^2) + nNg(vcosa)^2(\Delta y/\Delta t)(1-lny)/2Rg(c^2)(y^2) = 0$,where A stands for acceleration

⇨ $nNgvcosa(-(Acosa/1+lnv) + ((vsina)^2/Rg(1+lncosa)) + ((vcosa)^2(1-lnRg)/2Rg + (vcosa(\Delta y/\Delta t)(1-lny)/y)/Rg(c^2)y = nNg(1-lnRg)vcosa/Rg^2$ (solving this we will et acceleration in 3*10^8 meters per second square ,if we want it in meter per second square we must multiply L.H.S. by c ,we get

⇨ $nN_g v\cos a(-(A\cos a/1+\ln v) +((v\sin a)^2/R_g(1+\ln\cos a))+((v\cos a)^2(1-\ln R_g)/2R_g+(v\cos a(\Delta y/\Delta t)(1-\ln y)/y))/R_g(c)y = nN_g(1-\ln R_g)v\cos a/R_g^2$

⇨ $-A\cos a/(1+\ln v)+ (v\sin a)^2/R_g(1+\ln\cos a)+ (v\cos a)^2(1-\ln R_g)/2R_g+ v\cos a(\Delta y/\Delta t)(1-\ln y)/y = c(1-\ln R_g)(y)/R_g$

⇨ Now let us assume that our chosen galaxy is an average spiral galaxy. Scientists expect it to be between 0.02 and 0.08, and here I make my first prediction an average spiral galaxy will have an eccentricity of 0.05. Average $\cos a$ will be equal to $(e^2)/\pi$ for ellipses close to a circle, or, $(0.05)^2/3.14$

Or, 0.0008 apprx.

Now $\ln(\cos a)=\ln(0.0008) = -7.2$ apprx.

Now considering the star is at a distance of $3*(10^{20})$ meters away from the center of the galaxy (which is the distance between Sun and the center of milky way) $\ln(R_g)$ will become

$\ln(3*10^{20} *1.2*10^{25})$,$=\ln(3.6*10^{35})=105$ apprx.

Now, $-A\cos a/(1+\ln v)+(v\sin a)^2/R_g(1+\ln\cos a)+(v\cos a)^2(1-\ln R_g)/2R_g =c(y)(1-\ln R_g)/R_g$,ignoring the Δy term as it would be very small ,for now we can also ignore $(v\cos a)^2$ term as $\cos a$ is very small.

Or, $-A\cos a/(1+\ln v) +(v\sin a)^2/R_g(1-7.2) = c(1-105)/R_g$,ignoring y as it's very close to 1

Or, $-A\cos a/(1+\ln v) =c(-104)/R_g -(v\sin a)^2/-6.2R_g$

For acceleration to be zero

$$104c/Rg=(vsina)^2/6.2Rg$$

Since sina is soo close to 1 we may ignore it, then we have

$$v^2=6.2*104c$$

Or, $v^2=632c$

Or, $v=\sqrt{632c}$

Or, $v=4.354 *10^5$ meters per second

This is the velocity at which stars will revolve around the center of an isolated galaxy provided it's an average spiral one.

Now to analyze our equation,

⇨ -Acosa/(1+lnv)+ (vsina)^2/Rg(1+lncosa)+ (vcosa)^2(1-lnRg)/2Rg+ vcosa(Δy/Δt)(1-lny)/y= c(1-lnRg)(y)/Rg

vcosa(Δy/Δt)(1-lny)/y can be simplified to (v^2)Acosa(1-ln(c^2-v^2))/2(c^2-v^2)

For $v^2<<c^2$ we may ignore this term.

⇨ -Acosa/(1+lnv)+ (vsina)^2/Rg(1+lncosa)+ (vcosa)^2(1-lnRg)/2Rg+ = c(1-lnRg)(y)/Rg

Now when cosa>>sina we may ignore (vsina)^2/Rg(1+lncosa)

⇨ -Acosa/(1+lnv)+ (vcosa)^2(1-lnRg)/2Rg = c(1-lnRg)(y)/Rg

⇨ -Acosa/(1+lnv)= - (vcosa)^2(1-lnRg)/2Rg+ c(1-lnRg)(y)/Rg

For (vcosa)^2/2 >c acceleration will be in opposite direction or away from the center of galaxy, the critical velocity for outward acceleration of an object from a galaxy will be √2c or 24.5 km/s directly away from the center of galaxy. If cosa were 1/3 or a body were moving at an angle of 70.53 degrees away from center of galaxy ,the critical velocity would become 73.5 km/s and this is very much plausible if we assume a body generally moves only around the center of galaxy and gets a cosine velocity only due to chance (in spiral galaxy even the more).

If Acosa had been = -c/Rg +(vcosa)^2/2Rg then the velocity would increase proportionately with Rg or v/Rg=k, because of (1+lnv/c)(1-lnRg) velocity increase will be slightly greater or v/Rg=k+x, where x will increase very slowly.

Now, acceleration is dependent on (vcosa)^2/Rg ,and cosa approaches 1 as a body moves away from a reference frame so if cosa were less than 1 initially as the body started accelerating away from the center of galaxy cosa kept on increasing and so did acceleration and this increase will be much greater than the increase due to (1-lnRg)(1+lnv/c).

Now when A and cosa are very close to zero.

(1+lnv/c)(vsina)^2/Rg(1+lncosa) =c(y)(1-lnRg)(1+lnv/c)/Rg

Or, (vsina)^2=c(y)(1-lnRg)lncosa

Now this lncosa component is very important, because when cosa changes vsina changes by a huge amount due to this lncosa component and this is what happens in elliptical galaxies and star over there do not have a fixed velocity because lncosa keeps on changing. Now, before we come to the case of a planet revolving around a star in this galaxy let me remind you, differential of x is $\Delta x/\Delta t(1+\ln x)$ where x is an independent variable, if x is of the form a+b+c then differential will be $\Delta a/\Delta t(1+\ln(a+b+c))+ \Delta b/\Delta t(1+\ln(a+b+c))+ \Delta c/\Delta t(1+\ln(a+b+c))$ provided a,b and c are not independent of each other. In case of a body moving with velocity v1 ,its velocity is same with respect to all bodies at rest or with respect to all bodies moving at velocity v2 however it's velocity with respect to bodies at rest is different from the one with respect to bodies moving at v2. Therefore velocity with respect to one group of inertial objects (all objects moving at one velocity) is one variable, velocity with respect to another group of inertial objects is another variable. For Rg, for changing your distance from one object while while keeping it constant with respect to another, you must move in a curved path and that is not allowed as it will result in equivalent acceleration ,so when you change your distance with respect to one object, you change your distance from every object.

Now let us come to the case of a planet revolving around our star.

Suppose the planet moves towards the center of galaxy with velocity v1cosa and then it also starts moving towards the star with velocity v2 and the star makes an angle a+b from the direction of motion of the planet then the velocity towards the center of galaxy also increases or decreases and angle from the center of galaxy to the direction of motion changes from a to c.

We have

$$N(v2)^2/2(y)R + Ng((v1+v2\cos(a+b))\cos c)^2/2(y)Rg = Nc/R + Ngc/Rg$$

Now, $N(v2)^2/2(y)R$ is very small compared to $Ng((v1+v2\cos(a+b))\cos c)^2/2(y)Rg$,even for a planet as close as mercury to a star as heavy as the sun N/R is 100 times smaller than Ng/Rg so we may ignore $N(v2)^2/2(y)R$ in L.H.S. we have,

$$Ng((v1+v2\cos(a+b))\cos c)^2/2(y)Rg = Nc/R + Ngc/Rg$$

Now v1+v2cos(a+b) can be written as V and we may ignore y

we have,

$$Ng(V\cos c)^2/2Rg = Nc/R + Ngc/Rg$$

Differentiating both sides we get,

$$NgV\Delta V\cos c^2/Rg(1+lnV/c) - NgV^3\sin c^2\cos c/Rg(1+ln\cos c) = -N(1-lnRg)V\cos(a+c)/R^2 - Ngc(1-lnRg)V\cos c/Rg^2$$

We ignored some terms which have already been explained earlier

Or, $\Delta V\cos c/(1+lnV/c) - V^2\sin c^2/Rg(1+ln\cos c)) = -cRgN(1-lnRg)\cos(a+c)/NgR^2\cos c - c(1-lnRg)/Rg$

Now acceleration due to the star is

$$\Delta V\cos c/(1+lnV/c) = -cRgN(1-lnRg)\cos(a+c)/NgR^2\cos c$$

V was (v2cos(a+b)+v1)

$$\Delta(v2\cos(a+b)+v1)\cos c = -cRgN(1-lnRg)\cos(a+c)(1+lnV/c)/NgR^2\cos c$$

(A2cos(a+b)+A1-v2^2(sin(a+b))^2/R)cosc=-cRgN(1-lnRg)cos(a+c)(1+lnV/c)/NgR^2cosc

Or, Acosc - v2^2(sin(a+b)^2)cosc/R=-cRgN(1-lnRg)cos(a+c)(1+lnV/c)/NgR^2cosc

The cos(a+c)/cosc in the R.H.S. is because we are measuring the direction with respect to the center of the galaxy.

The equation can be simplified into

A - v2^2(sin(a+b)^2)/R=-cRgN(1-lnRg)(1+lnV/c)/NgR^2

Solve this and you will get double the gravitational acceleration from any object, double because we have taken an isolated galaxy. Now what if the galaxy isn't an isolated galaxy but part of a galaxy cluster and super cluster. Let's take Milky Way for instance, the closest large galaxy to it is the Andromeda galaxy but it will have a small effect as Ng/Rg due to it will be more than 1 magnitude lesser than Ng/Rg due to the center of Milky Way, however the virgo cluster situated at the center of our supercluster consisting of 2500 galaxies situated at a distance of 6.5*10^23 meters (2.75*10^3 times the distance between earth and center of galaxy) will exert considerable (almost 0.91% of Ng/Rg due to milky way) influence, as milky way is part of a weak cluster we may ignore the local cluster. The acceleration of sun towards the center of milky way will be obtained by solving

1.9Ng(vcosa)^2/2c(y)Rg=Ng/Rg

Or, vsina =435.4/1.9 km/s

=229.16 km/s

Now solving Rgc(1-lnRg)(1+lnv)/1.9Ng you will get the exact value of gravitational constant. (1+lnv/c) =ln229000/3*10^8 +1=-7.2+1=-6.2.the plane of solar system is mostly perpendicular(at an angle of more than 60 degrees) to the plane of the galaxy, so the velocity of earth around the galaxy won't exceed 232 km/s ,therefore (1+lnv/c) will be -6.2 for earth also.

Now,

1.9Ng(vcosa)^2/2√(c^2-v^2)Rg=Ng/Rg+N/R

Differentiating v^2/2√(c^2-v^2) we get,

vA/√(c^2-v^2)(1+lnv) + (v^3)A(1-ln(c^2-v^2))/2√((c^2-v^2))(c^2-v^2)

Over here while measuring ln(1+lnv) and ln(c^2-v^2) we must not forget c is one unit.

=-vA/√(c^2-v^2)6.2 + vAv^2(1-ln(c^2-v^2))/√(c^2-v^2)2(c^2-v^2)

= -(vA/√(c^2-v^2)6.2)(1-3.1v^2/c^2-v^2)

This -3.1v^2/c^2-v^2 will give the precession of planets like mercury. It will be slightly greater than relativistic values.(about 1 in 30 parts)

Till now we assumed all particles have equal and unit mass but this is not true for photons whose mass approaches 0, let m be the mass of a particle then its score is

Ng/Rg −mNg(vcosa)^2/2Rg(c^2)(y)

A photon's velocity approaches c so y approaches 0, so does m so they cancel out.

Ng/Rg −Ng(vcosa)^2/2Rg(c^2)

A photon has another component in its score called frequency which is something like the number of particles. Frequency can only change discreetly and therefore It's differential won't be divided by 1+Inf. But frequency change will make the score uneven as it will happen in an instant while other variables are changing continuously. Over here comes the concept of wave, score is unbalanced inside a wave, suppose there is a frequency change of 10 units in one go, then the photon will enter a wave till continuous change of other variables has fulfilled the score gap and then only the next frequency change will happen.

Score of a photon −

fNg/Rg - fNg(ccosa)^2/2Rgc^2

Now we want acceleration equivalent to be in m/s^2 and angle changes in radians so we multiply ccosa term with c.

fNg/Rg - fNg(ccosa)^2/2Rgc

or, fNg/Rg (1-c(cosa)^2/2)

The score of a photon is not balanced or zero, that is, it provides different amount of accelerations (or decelerations) to particles by being fired at different directions with different frequency. While being created the photon gives away it's unbalanced score so the change in the remaining score has to be zero to keep total universe's score change zero

Differentiating the score of a photon we get

$df/dt. Ng/Rg (1-c(cosa)^2/2) -fNgccosa(1-c(cosa)^2/2)(1-lnRg)/Rg^2 +fNg/Rg (-A(cosa)^2/2) +Ng c^2cosa(sina)^2/(1+lncosa)Rg^2=0$

$df/dt. Ng/Rg (1-c(cosa)^2/2) +Ng c^2cosa(sina)^2/(1+lncosa)Rg^2- fNg(ccosa)^2(1-lnRg)/2Rg^2-fNg/Rg(A(cosa)^2/2) = fNg(ccosa)(1-lnRg)/Rg^2$

Now for the moment let us ignore acceleration term and $fNg(ccosa)^2(1-lnRg)/2Rg^2$.

Then we have

$df/dt. Ng/Rg (1-c(cosa)^2/2) +Ng c^2cosa(sina)^2/(1+lncosa)Rg^2= fNg(ccosa)(1-lnRg)/Rg^2$

Now whenever cosa is naturally changing, the change will be reduced to compensate for right hand side term which you may call gravitation. When cosa isn't changing the frequency will change to compensate for R.H.S. term.

Case 1 : A photon directly moving towards the center of galaxy.

$df/dt. Ng/Rg (1-c(cosa)^2/2)= fNg(ccosa)(1-lnRg)/Rg^2$

$df/dt=fccosa(1-lnRg)/Rg(1-c(cosa)^2/2)$

let us approximate the denominator as $-c(cosa)^2/2$ then,

$df/dt = -2fc\cos a(1-\ln Rg)/Rgc(\cos a)^2$

$df/dt = -2f(1-\ln Rg)/Rg\cos a$

For a small Δh corresponding frequency change will be

$\Delta f = -2f\Delta h(1-\ln Rg)/c\cos a Rg\cos a$

Case 2: Photon is moving directly towards a massive body.

$df/dt. Ng/Rg (1-c(\cos a)^2/2) = 2fN(1-\ln Rg)c\cos b/R^2$

$df/dt = 2fNc\cos b(1-\ln Rg)Rg/Ng(1-c(\cos a)^2/2)R^2$

let's us approximate $1-c(\cos a)^2$ as $-c(\cos a)^2$ and $\cos b = 1$

$df/dt = -2fN(1-\ln Rg)Rg/Ngc(\cos a)^2(R)^2$

we know $G = -Rgc(1-\ln Rg)6.2/2Ng$

$df/dt = 4fNG/6.2(c\cos a)^2(R)^2$

$= 2fNG/3.1(c\cos a)^2(R)^2$

This is the change in frequency due to gravity.

Case 4: A photon moves towards a star grazes it's surface then passes away.

In this case cosa is naturally changing and total cosa change will decrease such that there will seem to be a deflection, there will be no frequency change or gravitational redshift in this case as deflection will compensate for the required score.

$fNgc\cos a\Delta(\cos a)/\Delta t(1+\ln\cos a)Rg = fNc\cos b(1-\ln Rg)/R^2$

$\Delta(\cos a)/\Delta t(1+\ln\cos a) = N\cos b(1-\ln Rg)Rg/Ng\cos a(R)^2$

Now, when N/R is much smaller than Ng/Rg no significant change will occur , significant changes will occur only when N/R is comparable to Ng/Rg, there will come a time when N/R will be very close or even greater than Ng/Rg from center of galaxy, we may write total Ng/Rg as (Ng/Rg + N/R). To further simplify our calculations we will take R/N instead of Rg/Ng and multiply the whole by N/R*Rg/Ng.

$$\Delta(cosa)/\Delta t(1+lncosa)=Ncosb(1-lnRg)R/Ncosa(R)^2 * N/R * Rg/Ng$$

$$\Delta(cosa)/\Delta t= (1+lncosa)Ncosb(1-lnRg)R/Ncosa(R)^2 * N/R * Rg/Ng$$

Since total Ng/Rg= N/R + Ng/Rg

Rg/Ng= RRg/NRg+NgR

$$\Delta(cosa)/\Delta t= (1+lncosa)Ncosb(1-lnRg)R/Ncosa(R)^2 * N/R * RRg/NRg+NgR$$

$$\Delta(cosa)/\Delta t= (1+lncosa)cosb(1-lnRg)/cosaR * NRg/NRg+NgR$$

NRg=6*10^50 For N = mass of our Sun

NgR=2*10^50 for R=6.95*10^8

Now let us integrate the instantaneous change in cosa, we get approximately

$$\Delta(cosa)= \Delta ln(R)(1-lnRg)(1+lncosa)cosb/c(cosa)^2 * NRg/NRg+NgR$$

Now let us assume N/R = total Ng/Rg we will later multiply the value we get by the fraction N/R was of total Ng/Rg during the whole journey ,then cosa becomes cosb

$$\Delta(cosa)= \Delta ln(R)(1-lnRg)(1+lncosb)cosb/c(cosb)^2$$

This is the value we get if photon moves towards the center of galaxy (with cosb being the angle it makes with the center of galaxy)

Now suppose the photon started travelling from hades group of stars (as was used in famous experiment of 1919) 10^18 meters away from the Sun and grazes the Sun with a radius of 7*10^8 meters and reaches earth 10^13 meters away. ΔR=10^18 - 7*10^8 + 10^13 meters, now we must take Δ ln(R) in solar units.

Δ ln(R)=ln(1.428*10^4)-ln(1) –(ln(1) –ln(1.428*10^9))

Δ ln(R)=ln(1.428*10^4)+ln(1.428*10^9)

Δln(R)=ln(2*10^13)

$\qquad$ =30.5

(1-lnRg)=104

Cos b changes from -1 to 1 so (1+lncosb)=1

Δ(cosa)= Δ ln(R)(1-lnRg)(1+lncosb)cosb/c(cosb)^2

Δ(cosa)=30.5*104/3*10^8

=2.15 arc seconds

This would be the deflection if the photon grazes the center of galaxy travelling equivalent distance

For our Sun the deflection would be

2.15*∫dx , (where x = NRg/NgR+NRg)

Now x changes from a very minuscule quantity to 6*10^50/1.4*10^50+6*10^50 at cosb=0 and back to a very small quantity at cosb=1,(mass of center of galaxy is 2*10^41 as its measured to be 10^42 including dark matter which does not exist)

We have

2.15*(6/7.4)

=1.743 Arc seconds

As you can see from the equation this should be a somewhat longer range effect than Einstein thought, meaning that for R>7*10^8 this deflection should decrease lesser than he thought.

Now coming to the other 2 terms in photon's score

fNgACosa/Rg=-fNg(ccosa)^2(1-lnRg)/Rg^2

Now photon's velocity is always c, how can it accelerate, well it doesn't. Look at it this way, for a particular score change if you could accelerate a particle to velocity v, at twice that much distance from the center of galaxy for same amount of score change you can accelerate that same particle to more than 2v. Time is approximately equal to Rg/Ng(y), such that velocity of light is always constant. As photon moves away from center of galaxy its velocity decreases due to time dilation,because of Acosa= (ccosa)^2(1-lnRg)/Rg, the velocity remains c.

Momentum and Kinetic Energy

A photon increases the number of particles in the universe, therefore when a photon is released all the particles in the universe will experience a score deccrease of f(ccosa)^2/2cR, there will be a total score decrease of Ngf(ccosa)^2/2cRg , a photon carries an unbalanced score of -Ngf((ccosa)^2 -2c)/2cRg, (which is immediately given away at the time of creation) so the net change in score of the universe will be $-fNg(c\cos a)^2/2cRg$ from the creation of a photon. So a photon release should always be accompanied by a net velocity decrease in the universe and photon acceptance by an increase.

$-(v\sin a)^2/Rg(1+\ln\cos a) - (v\cos a)^2(1-\ln Rg)/Rg +A\cos a/(1+\ln v) = -c(1-\ln Rg)/Rg$

vsina term and acceleration term will always sum up to the c term unless Acosa changes due to a significant change in Ng/Rg ,if vcosa changes due to orbital perbutations, vsina or unbalanced acceleration will increase.

$-(v\sin a)^2/Rg(1+\ln\cos a) +A\cos a/(1+\ln v) = -c(1-\ln Rg)/Rg$ unless Acosa changes due to change in Ng/Rg. $- (v\cos a)^2(1-\ln Rg)/Rg$ ensures that vcosa/time or Ngvcosa/Rg(y) remain constant (a little more than constant due to new calculus) .

$Ngv\cos a/Rg(y) -(v\sin a)^2/Rg(1+\ln\cos a) +A\cos a/(1+\ln v)=-c(1-\ln Rg)/Rg +k$

$Ngv\cos a/Rg(y)=k$

For n number of particles $nNgv\cos a/Rg(y)=nk$

Now if this body of n particles collides with another body of n2 particles moving at v2 velocity

$n2Ngv2\cos a/Rg(y)=n2k2$

Now change in nk must be equal and opposite to change in n2k2 else score will change,

$\Delta\, nNgv\cos a/Rg(y)= \Delta n2Ngv2\cos a/Rg(y)$

Δ nvcosa /(y)= Δn2v2cosa/(y)

This gives us the conservation of momentum. For the change in kinetic energy photons will be released.

EMF

Whenever a charged particle spots another charged particle it releases a photon inside itself which is again accepted inside itself, this way no extra particle is added into the universe but the unbalanced score of the photon will have to be accounted for and it will be by acceleration towards or away from the other charged particle.

Unbalanced score of a photon = -fNg(c(cosa)^2-2)/2Rg=Ng/cRRg or – 1/cR where R is our chosen particle's distance from the other charged particle .

f(c(cosa)^2-2)/2=1/cR or -1/cR

-fNg(c(cosa)^2-2)/2Rg will cause a (vcosa)^2 decrease of f(c*c(cosa)^2-2)/2

Δ(vcosa)^2 should be = f((ccosa)^2-2)/2

But v is a continuously changing variable because of gravity so the discrete release of photons will not affect (1+Inv) but velocity is a continuously changing variable and any change in it should always be divided by $(1+Inv_o)$ where v_o is the velocity due to continuous acceleration such as gravitation.

Δ(vcosa)^2/(1+Inv_o) = f((ccosa)^2-2)/2

Δ(vcosa)^2/(1+Inv_o) = 1/R

Now vcosa will have a discrete change and it will enter a wave where R will change continuously till the score gap is filled.

$\Delta(vcosa)^2/\Delta t(1+Inv_o) = \Delta(1/R)/\Delta t$

$2vcosaAcosa/(1+Inv_o) = (1-InRg)vcosa/R^2$

$Acosa = (1-InRg)(1+Inv_o)/2(R^2)$

For R=1meter the acceleration of a electron towards a positron will be

$Acosa = -104*6.2/2$

$= -322.4 \ m/s^2$

This is very close to the experimental value of acceleration of electron towards positron. Now a proton's acceleration towards an electron, in a proton only one particle does the attraction probably a positron inside it, the rest of the particles composing the proton are ordinary particles.

Proton's acceleration towards electron:

$n\Delta(vcosa)^2/\Delta t(1+Inv_o) = \Delta(1/R)\Delta t$

Where n is the number of particles in the proton into their respective mass, which is equal to mass of proton/mass of electron =1837

$Acosa = 322.4/1837 \ m/s^2$ for R = 1

Because acceleration from photons is instantaneous and R change continuous there will be a component of vsina in R change. R after a given time t will be $R+vcosa(t)+\sqrt{(R^2+(vsina^2)*t)} -R$

$= R + vcosa(t) + (vsina^2)*t/2R$

$nAv(cosa)^2 = -(1-lnRg)(vcosa)/R^2 - (1-lnRg)(vsina^2)/R^3$

$nAcosa = -(1-lnRg)/R^2 - (1-lnRg)v(tana)(sina)/R^3$

The second term in R.H.S. is for magnetism.

The End